Eugenio Gallavotti

Everything we always wanted to know about

SUICIDE

(Because it is not something
that can happen to anyone)

Mario Savino

3

PAMPHLET/NON-FICTION
Reading time 1 hour

Graphic Design: Alessandro Laganà
Font: HK Grotesk
2019

Codice ISBN: 978-1-79478-136-8

Content

This publication discusses suicide, which is a topic that may be frightening so we reject it, but the tone is one of hope, inspired by new scientific research and new interpretations. Our intention is to debunk the many clichés heard, all too often repeated even by the most prestigious press, namely that suicide is an act of heroism or of cowardice, that it is "incomprehensible" or "mysterious", and so on and so forth.

There's no mystery: in the majority of cases, suicide is simply the most dramatic epilogue of a widespread ailment – depression – that is sometimes underestimated or unrecognized even by its sufferers. It's incredible that in the third millennium we still speak of "mystery" when 2,400 years ago, Hippocrates was already writing about "melancholy".

One of our key concepts offers hope: suicide is NOT something that can happen to everyone. Our brains have inhibition mechanisms for self-protection that in most of us are steady and solid. The latest studies from the University of Utah even suggest a "suicide gene". In short, those who aren't born with the gene will not take their own lives, whatever negative event they face.

The publication offers hope to millions of people suffering from depression, and for their families and friends, describing new treatments and techniques for overcoming mental disorders.

The publication offers hope because it is auto-biographical... The author entered the darkest of tunnels on 18 June 2002; the coauthor of the book, Professor Mario Savino (who stud-

ied with Giovanni B. Cassano in Pisa), is the psychiatrist who helped me see the light at the end of the tunnel. Yes, indeed it does take a long time to come to terms with yourself.

The text comprises several sections: the predisposition to suicide; new treatments for depression; suicide as reported by the media; suicide in women, in the young and the very young; in animals; the social danger of potential suicides; the successes and failures of a psychiatrist, as told by Mario Savino; assisted suicide; appendices dedicated to brain stimulation machines, psychotherapy, and prevention.

Suicide is in the genes and if the carrier meets a hostile environment, the worst will happen. Diagnosis and appropriate remedies, on the other hand, can provide a cure. Yet there is still a surprising lack of awareness on the subject, even among "educated" people. For example, when someone comments: "But why did he kill himself? He had everything" or "Suicide? An act of weakness." Well, those who really are weak can't kill themselves. Indeed, "people full of life" are the most at risk: if they get sick, they may end up in the inferno of the so-called "mixed state", which is a depression combined with anxiety and agitation. And this is the most dangerous state because that person has all the physical and mental energy to end it all.

E.G.

PART 1

Suicide
is genetic

Why doesn't the African mother who sees her children drown in the Mediterranean let herself die with them, while the top manager who has everything – family, career, money, rewards of all sorts – opens his office window one day and jumps out?

It's a topic as complex as it is simple. We can start with studies conducted by psychiatrist Mario Savino in collaboration with the University of San Diego. The most frequent pathologies in certain professional categories were analysed and it was discovered that the same character traits are found in a number of professions. Engineers, for example, are usually very precise, attentive to detail but they don't necessarily suffer from any obsessive disorder although they tend to that type. And there's no getting away from the fact that if you don't have those traits, you can't be a really good

engineer... The same goes for artists, architects, journalists, creatives in general, those
who show initiative and have that extra something, but they pay a price because when
they're at their most prolific and productive,
they will be exhausted.

If these subjects fall ill, any depressive crisis
is often accompanied by agitation. This is the
most dangerous type of depression, because
the sufferer is "slowed down", "blocked", and
they don't even despair, they just sit on the
sofa, stop eating, think of suicide but at best
will only have the strength to indulge in a few
attempts at self-harm (remember Mel Gibson
in The Beaver?). In short, they are less at risk.

Now, if the African mother doesn't show these characteristics, it's more likely she'll survive her terrible pain, perhaps devoting herself to her other children or her grandchildren, while the "hyper", energetic person, paradoxically, is more vulnerable.

So, suicide is in our DNA? Is it scientific to say that human beings are divided into those who have the "suicide gene" and those who don't? Are some of us predisposed to suicide while others would never dream of it, even when faced with the most atrocious events, the most insidious sources of stress?

Recently, researchers at the University of Utah Health identified as many as four genetic changes that occur most frequently in suicides and these changes indicate an increased risk in vulnerable individuals.

Suicide is the tenth highest cause of death in the United States and ranks alongside the number of deaths caused by opioids. Previous studies show that suicide creeps into families regardless of the effects of a shared environment. American researchers used extensive resources to identify genetic factors that may increase a suicide risk and results are available online in the Molecular Psychiatry journal.

"Research on families and twins revealed there's a significant genetic risk associated to suicide," says Douglas Gray, a University of Utah Health psychiatry professor. "Finding the genes that increase the risk could lead to new treatments."

The team was able to identify variants in four genes (SP110, AGBL2, SUCLA2, APH1B) that might increase the risk of death by suicide. The variants were not present in individuals who died from other causes.

Suicides in forty-three families were examined and the conclusion was that the genetic risks of suicide downscaled the effects of a shared environment, such as stress linked to divorce or unemployment, or easy access to lethal means.

Researchers looked at genetic variation in more than 1,300 DNA samples from people who died by suicide in the state of Utah. They identified specific changes in four genes, but also two hundred and seven genes warranting further analysis to understand the role they played for people who committed suicide.

Of these genes, eighteen were previously associated with suicide risk. Of the previously identified genes, another fifteen were also associated with inflammatory conditions, upholding increasing evidence on the relationship between inflammation and mental health.

University of Utah Health researchers concluded that suicide is no different to any complex human condition. There may be a variety of genetic factors making us more prone to risk and, of course, there are several environmental factors that can intensify it.

In short, suicide is genetic and paired with negative environmental factors, a risk becomes a fact.

We no longer want to quote Freud, who thought suicide was always an act of revenge against a part of ourselves we can't deal with. It's an outdated definition.
There's now no doubt that some of us are unlucky enough to be born with a gene that is often hereditary, which carries a mood disorder and its inherent risk, which can generate acts of self-harm.

Most of the time, the tendency to suicide affects people who suffer from mood disorders and are biologically predisposed to the act. Alterations in the functioning of serotonin and other neurotransmitters have been proven and are evident in subjects who commit suicide but aren't found in those who die from other causes.

The link between chemistry and suicide is now recognized: in our brains there are inhibition mechanisms for self-protection but they may not be quite so solid and steady for everyone. So, even to suggest suicide is an act of cowardice or of heroism are now very much outdated concepts. Very outdated indeed.

Many have tried to explain, to interpret the abyss that engulfs those who voluntarily "alight from life". An action beyond ken, psychological problems, perhaps of a biological nature... In reality, today we're able to sketch an identikit of a potential suicide, "based on data," says Professor Mario Savino.

Suicide is more common in those who suffer mood disorders, chronic anxiety disorders (especially if associated to mood), eating disorders, personality disorders, in abuse of substances like drugs and alcohol… And here we might linger on the so-called "mixed state depression", thus defined and studied by the most experienced psychiatrists.

The "mixed state" is not a form of pure depression, as many know or imagine it, namely the profound depressive state often called "major" for its severity, which "blocks" or "slows down" the sufferer. This condition does not carry great risks, except at start of therapy. On the other hand, it is above all bipolar depression that leads to suicide and appears in three different guises: pure depression; euphoric/manic depression; mixed state depression, a sort of combination of the other two. Hence the mood can be depressed or swing from one pole to the other, accompanied by agitation, particularly difficult anxiety, and sometimes irritability and aggression.

Thus, the potential suicide is not oppressed by "sadness", but rather by "agitation", by an unstable, uncontrollable condition, which can generate violent behaviour towards oneself and others. The subject is ill but at the same time strong and active, which is the most dangerous condition because it veers to impulse actions. The mixed state may arise spontaneously or as a result of wrong diagnosis: for example, if a patient in a mixed state is treated with antidepressants, without effective stabilizers, risk grows.

What's the best thing to do then? Seek a precise diagnosis, and not only based on how the psychiatrist "sees" the subject at that time, but also based on clinical history, including that of the family. For example, a patient who has family members suffering anxiety disorders and alcoholism, and appears depressed, could be on the bipolar spectrum. Often what is considered a major unipolar depression in early diagnosis is then found to be a hereditary bipolar disorder (as a predisposition, but the environment also has an impact), especially among juvenile-onset patients and when there are frequent recurrent depressive episodes, for example, women suffering severe postnatal depression.

E.G.

Professor Savino, when a person decides to commit suicide in a relatively short timeframe, does psychotherapy become superfluous?

M.S.

The answer is "yes". It's true that psychotherapy has changed a lot, new techniques exist, and some are very important. But psychotherapy that says "Let's start from your childhood and explore the developments, without medication, because you need to work through it yourself," is a waste of time in an emergency, because the patient is so ill they won't benefit significantly from that technique.

"Nobody commits suicide because they want to die."

Tiffanie DeBartolo

E.G.

Emergencies are something that every psychiatrist has experienced firsthand...

M.S.

Fortunately, touch wood, in almost thirty
years, very few of my patients have
been suicides.

E.G.

We'll go back to that. But meanwhile let's fin-
ish talking about causes and interpretations.
Until yesterday, suicide seemed mostly relat-
ed to a variety of psychological factors...

M.S.

Today we can say that suicide can be prevented only if the diagnosis is correct (mixed state or agitated depression), regardless of the cause of distress. If the patient responds to stress with a slowed depression, in other words basically takes to their bed, there is less risk; if they respond with anxiety and agitation, the risk is much greater. And if someone is considering suicide, going back over the story of their life is pointless: they need lithium or electroshock treatment... It's too late for talking it over.

PART 2

The revenge of electro-shock

What has scientific research achieved so far in the eradication of depression, an illness that is still underestimated today, indeed, sometimes it isn't even considered a pathology?

In many countries worldwide, including Italy, depression is second only to cardiovascular disease as a cause of social and professional disability. There is clinical depression, which is neither sadness nor discouragement, but an unbearable and hopeless anguish. It's an illness that runs in families and, above all, it's inherited. There are good antidepressant medications and good techniques available...

On 18 June 2002, after Korea defeated Italy in the World Cup, I sank into the most hopeless time of my life. Not because we lost the match, of course, but following my divorce, I quickly fell prey to a fierce dysphoria, turning into a sort of Dr. Jekyll and Mr. Hyde. Professor Mario Savino treated me with ECT – electroconvulsive therapy, better known as electroshock. Yet people are still very diffident of ECT...

As a young specialist, Savino used this technique and came into direct contact with the risks involved, but also with the benefits it brings. It was a system that lent itself to being stigmatized as state torture, because of huge prejudice against it. Well, I'll say loud and clear that it's one of the most effective therapies for beating suicide because it treats so-called mixed-state episodes, not just the depression but also the agitation. And it works very quickly.

We'll stop reading ridiculous online headlines like "Auschwitz-style torture" when the

new, lighter, simpler stimulation practices like magnetic or direct current (administered without anaesthetic) replace ECT.
Although it should be added that at the moment the new techniques are often slower to take effect.

Even primetime TV has spoken out against ECT. For instance, one show in Italy invited the guru of psychiatry Athanasios Koukopoulos (who died in 2013 aged eighty-three), a student of the Italian inventors of electroshock, which helped to overcome potentially deadly insulin coma simply by triggering a convulsion with a quick application of current. The anchorman almost insulted Professor Koukopoulos during the live show because he didn't want electroshock mentioned during his programme. An open display of his ignorance of the benefits of therapy.

Actually, Koukopoulos was right.

In a depressed patient who isn't respond-ing or is intolerant to medication, ECT is the only tool that can give relief without render-ing them euphoric or suicidal. Says Savino "Drugs used badly do much more damage, it's not easy to treat with drugs. It takes time, it's a process of trial and error. ECT has a few drawbacks (it's not appropriate for certain patients) but it certainly brings benefits. To-day, thanks to new unilateral ECT, there are hardly any risks of memory loss, and it is a small amount that is recovered later.

Let's face it, people talk as much nonsense as they want."

Depression affects many people, and their families, but there are new therapies, new strategies to curb it, which can be seen in cur-rent research and in years to come.

First of all, physicians today pay more attention to diagnoses that involve high risks. Unfortunately, these are sometimes are generically classified simply as depression and treated partially or in any case imperfectly according to the guidelines. In a few years, new drugs will be available to us. There are also stimulation therapies, such as transcranial magnetic therapy or direct stimulation, which are delivering promising results. Unlike ECT, modern electroshock, which is still the best. In truth, there are reports or articles saying ECT works because it creates amnesia and we forget our woes, so we heal. A highly questionable standpoint. The amnesia isn't so constant or deep, it doesn't last so long, nor is it so important. It usually arises during treatments. Then the patient, ECT notwithstanding, will have the same problems as before, but their mood improves thanks to the induced seizure.

Major depression is far more common these days than in the past. It can be bipolar or un-ipolar, the latter not alternating with mania, excitement, agitation, hyperactivity.

In recent years, the percentage of bipolar depression has also risen, in part because psychiatrists have become more skilled in identifying its less typical forms. It's no longer a question of small numbers but of an import-ant slice of cases. Going beyond the basic examination and drawing up a more accurate case history to pinpoint certain indicators, also recruiting the help of family members, a bipolar depression can be diagnosed even if the subject has never shown manic episodes. In this way, we don't run the risk of mak-ing things worse with antidepressants, which should be used carefully because if the diag-nosis is not correct, the patient deteriorates.

It has been said that antidepressants are a shot in the dark, difficult to get on target... There's no doubt they are precious weapons, to be handled with care. Any drug, if it is the wrong drug, is either useless or aggravates the disorder. Antidepressants have been used too much and for too long in those depressions where they shouldn't have been used at all or should have been associated with a mood stabilizer or antipsychotic medication to slow down agitation or anxiety factors. Tricyclic drugs also work, although they are a little dated now...

The problem is that in some countries, the excellent mood stabilizer lithium is prescribed less often than it should be. This suggests that some professionals fail to research the drug-stabilizer combination. In theory, ideal bipolar depression therapy should be entrusted only to stabilizers; in practice, there are situations where a depressive phase lingers or is particularly serious, and the patient therefore needs to be helped and to give them relief, antidepressants deemed less likely than others to induce agitation are prescribed.

One serious mistake is self-treatment, with or without medication. It is essential to ask for help, to engage in a relationship with a second person. In short, the depressed person has to want to heal and take care of themselves seriously. It's not easy, because depression is accompanied by a conviction of incurability, the certainty that no one will ever be able to help. Mario Savino says: "I have a friend who underwent regular treatment for two years, having suffered a serious episode. I didn't understand why he didn't improve. Then, when he was finally better, taking his medication, he confessed that for two years he hadn't taken it, leaving it in a drawer, and he just stayed in bed all day. Some people think it's "humiliating" to take medication, while for others it's a chore. In cases of major depression there's the idea that it's all useless, so there's no point in taking tablets. But an anxious subject suffers panic attacks and usually asks for help. My friend spent two years in the depths of despair and without telling anyone. Now he takes lithium. He's bipolar too, without ever having true manic episodes, though he's hyperactive, extremely outgoing and creative, one step ahead of many other people. But predisposed family genes combined with external environmental factors made him ill.

There are people who spontaneously stop treatment for depression, obviously when they feel better. But it's very dangerous to do that, especially with stabilizers: lithium, or anti-psychotics such as clozapine, are associated with a high percentage of relapse after people stop taking them. In practice, for seriously sick patients, they save lives. It is possible to decrease the dosage as patients often push to reduce drug therapy when they feel happier, gratified, healthier, but it's very risky to decrease or even stop outright.

The suicide input is sometimes the instinct of self-preservation."

Stanislaw Jerzy Lec

The first warning lights appear when we're about to run out of steam: restless sleep, loss of enthusiasm, reduced memory and concentration, fading interest in sex. In general, a drop in vital signs. Appetite may increase or decrease, so those are not reliable symptoms. Or, a person prefers to spend time at home alone watching TV, doesn't read any more, and maybe starts to suffer from headaches or other physical ailments.

Many patients report they felt well and then tumbled into the disease in just a few hours, even showing physical changes like a swollen leg or ankle. In other words, the brain is sick, but the rest of the body isn't joking either. Because depression is a systemic disease of the entire body, with alterations of various functions, including bowel, cognitive, sight, physical energy, endocrine and hormonal abnormalities. It's wrong to say that this is a mental illness: it involves almost the whole organism in a demonstrable manner.

So, when do most suicides occur? Generally, at the beginning of the pharmacological therapy. Lithium isn't yet "activating"; antidepressants act slowly and often with an initial deterioration. In addition, the depressed person may be feeling more proactive thanks to the treatment, while thought processes are still pessimistic and hopeless. In this situation, a suicide attempt is more likely. Therefore, the "First Aid Manual" suggests accurate diagnosis and prudent use of antidepressants, in addition to a firm choice of quickly effective treatments, such as so-called physical therapies that do not include the use of drugs or flank drugs with therapies implemented through brain stimulation. What we call electroshock, for example, is a physical therapy, and perhaps the first to be truly effective, although at the beginning it was a dangerous stimulation that produced seizures, using substances such as Cardiazol or insulin. Today, modern ECT brings on an epileptic seizure that is therapeutic in itself.

In autopsies of patients who underwent hundreds of ECT applications during their lifetime, no brain damage came to light, because – like all stimulation therapies and like drugs – these techniques induce a proliferation of neurons, increasing the number of connections. Other techniques are not yet as effective as ECT, but they do help in certain forms of drug-resistant mental disorders. They are based on magnetic stimulation which doesn't usually cause seizures. Other techniques include direct current brain stimulation (which

doesn't induce seizures either, and is pain-
less) and vagus nerve stimulation.

As we can see, there is a vast spectrum of
possible actions, which include functional
neurosurgery, the least invasive surgery of
this type possible, placing electrodes in precise areas of the brain, useful against depression but also against severe obsessive-com-
pulsive disorder resistant to medication and
psychotherapy, Parkinson's disease, and other neurological disorders.

The common denominator of all these techniques seems to be the idea of "jump-start-
ing" something, like thumping the television
sets of the past when they played up.

In reality, it's often not a case of "restarting"
but of inhibiting something. For example, in
deep intracranial stimulation, the electrodes

can usually inhibit nuclei of hyperactive cells causing a disturbance.
Sometimes we need to step on the brake rather than on the accelerator.

A complex of real or perceived guilt is one of the triggers of depression. In general, even without causing irreparable harm, if we interpret every event as our fault or when we blow up trivial errors out of proportion, the truth is that nothing serious has happened. In short, when we feel guilty of everything, from the fate of the world and everything in between, this is a clear sign of depression.

There is also the guilt felt by those close to a suicide, relatives and friends who say, "I could have done more". Here, too, we must bear in mind scientific experience, data. For some people it's genetically easier to take their own lives, usually – but not always – due to mood disorders. There are also those who have another suicide or attempted suicide in the family. So, some commit suicide not because they are sicker than someone else, or even through any fault of ours ("I could have done more"), but because often there's a predisposition that causes them to behave in a self-destructive way. The "guilt" felt by those close to the suicidal subject is not as central as we think. A strong factor comes from DNA, from the working of our brains, because there are those of us who are better protected by neurotransmitters that function more efficiently in some and less so in others. We shouldn't, therefore, add guilt to grief.

E.G.

Professor Savino, is it true that depression can also be triggered by a bout of flu?

M.S.

Depression, impaired immunity, inflammation and other bodily reactions have always been the subject of research.

Moreover, some time ago, studies into the possible causes of depression identified the most important factors like genetics, environmental aspects like long-term trauma and stress, endocrine disorders, but also contributing aspects, which act only in conjunction with favourable conditions, in highly susceptible individuals.

These contributing factors include inflammation (I always recommend my patients

get flu jabs to reduce the risk of relapse)
and it seems that it's precisely inflammation that makes stress trigger depression,
namely it's the fuel for stress to start a
depressive episode.

This new role played by inflammation has
new implications in the treatment and prevention of mood disorders.

E.G.
How does inflammation affect depression?

M.S.
Even a bout of flu can decrease serotonin
and noradrenaline, which are important
transmitters in maintaining proper mood
quality. In addition to this, inflammation
increases the level of cortisol, a stress
hormone usually found to be high in de-

pressed people. Nowadays, of the depression therapies proving to be effective, there are many that also have antiinflammatory effects. One example is the SSRI (Selective Serotonin Reuptake Inhibitor) antidepressant, but there also hypericum and Omega 3 fatty acids. Today we take it as given that inflammation is a key factor in depression and that psychophysical stress (the well-known depressogenic stressor) causes an inflammatory response of the organism, in turn capable of triggering depression.

E.G.
In short, do anti-inflammatories have an acknowledged role in psychiatry?

M.S.

Inflammation and depression seem to be closely linked. We might suppose that at least a part of depressed patients would also benefit from an anti-inflammatory treatment, or at least when the overview of the depression-inflammation relation-ship further clarifies all aspects.

PART 3

Women and young people

E.G.

Professor Savino, women commit suicide far less often than men. Is it because they have less physical strength? We have seen that it takes forcefulness to commit suicide...

M.S.

But women try to commit suicide more often than men do. Women try and fail. In part because their actions are less violent, like taking an overdose, and in part because a woman's affective life and reactivity to circumstances are more multifaceted than those of a man. Sometimes a female suicide attempt may be driven by recrimination or may be a cry for help.

E.G.

A new drug to treat post-natal depression was recently approved by the Food and Drug Administration, the US agency tasked with verifying safety of medications. Approximately fifteen percent of women experience mood swings in the weeks following childbirth...

M.S.

Yes, it's very interesting drug, a hormone and the first of the new medications we've been waiting for. New-generation drugs to give us a hand will be hormonal and anti-inflammatory. Ketamine seems to be on a downward curve because of the associated psychotic effects.

"Even though I was
a happy child,
I suffered.
And this suffering
was like a boulder."

Saoirse Kennedy

E.G.

Young people and suicide:are data worrying?

M.S.

Unfortunately, yes. In the young, suicide is the second highest cause of death after road accidents. There's a problem of alcohol abuse than has rocketed over recent years. Our kids' parties often end up with instances of alcohol-induced coma. It's not greater consumption but bouts of more concentrated drinking on certain occasions, reaching dangerous levels.

Thus, inhibition is removed, and subjects commit suicide who would never do such a thing if they were sober. Moreover, the young are more impulsive than adults, and often easily provoked by certain dynamics, like the infamous virtual challenges on

the internet. Alcohol and the Web make
for a deadly cocktail.

E.G.
Young people are more likely to commit
suicide than other age groups.
Is bullying also an issue?

M.S.
Abuse can lead to depression and anger,
and often to thoughts of death. Again, diagnosis counts: young people are at much
higher risk because of social phobia,
BDD, for easing about a physical defect
that makes them feel repugnant.

At a young age there can be triggers, even
some kinds of heavy metal. Recently, a
sixteen-year-old Malaysian girl committed suicide after posting an Instagram poll

asking her followers if she should commit suicide or not.

She took her own life after sixty-nine per-cent of users voted for suicide.

A research team from the National Institute of Mental Health recently found that near-ly a third of young people aged ten to twelve tested positive for suicide in US emergency departments. "Generally, suicidal thoughts and behaviour are noted in older children. It's worrying to see that so many preteens are considering such a terrible option," says Lisa Horowitz, a scientist at the institute.

"Where I grew up in Brooklyn, nobody committed suicide. Everyone was too unhappy."

Woody Allen

PART 4

Media, social hazards and assisted suicide

We often read of "mystery", of an "incomprehensible gesture", even in top newspa-
pers, in news reports of suicides. Subjects
who are "brilliant" and "hyperactive" and kill
themselves astonish all of us. We know that
precisely the mixed state, the agitation and
restlessness, the capacity to act, will render a
suppressed or inadequately treated depressive disorder dangerous. On this issue, an
almost mediaeval ignorance prevails, even in
more savvy social classes. Once and for all, it
has to be clear that there's no mystery. Suicide is the most tragic ending of a widespread
illness – depression. Today we must stop calling it a "mystery".

Yet journalists pay close attention to how to
give the news of a suicide, the language used,
how to approach the description these dis-
turbing episodes. In some cases, the news

is almost hidden, given a small box at the bottom of the page, as if embarrassment or fear had the upper hand. The World Health Organization regularly publishes a manual of guidelines for the communication industry. Does this approach somehow help save other lives? Can a refusal to "trumpet" the event be considered a form of prevention?

Here we must be very clear. We must avoid descriptions of some of the methods used by young people, those kids who are now risking their lives, "getting a kick" out of dicing with death, committing "partial suicides". Unfortunately, the Web provides plenty of "instructions" for these activities.

In addition, the WHO refers to suicide outside the family circle, which can trigger emulation, for instance in the case of famous figures. The private sphere is very different: here there is an unfounded conviction that talking about suicide is "contagious", that exploring the intentions of a victim means inciting those who already suffer from a psychiatric disorder. "On the contrary," says Savino, "some studies show that traumatizing events like the suicide of an acquaintance reduce the risk in the subject who learns the news".

The contrary of what we have always thought. Suicide, in these cases, becomes discouraging and even a warning. We start to think: "Too bad, they could have solved their problems in many other ways and now we can't do anything anymore… " It's a shock that gives pause if someone had been thinking of suicide. So, we shouldn't be cautious. It should be discussed openly, asking if there are suicidal thoughts, understanding why it's useful

for someone who's struggling, because it per-
suades them to ask for help, which isn't easy.

E.G.

Is a potential suicide also a danger to society?
Professor Savino, remember the cases of pi-
lots who used their planes to commit suicide?
We say a suicide harms only the person who
takes their own life. Now, it seems to me that
the discussion on prevention is also important
because the gesture truly can involve other
people. A suicide is not necessarily innocuous
for others.

M.S.

A suicide is also a danger to society.
The mother who suffers from post-natal
depression constantly thinks about how
her child will survive when she's dead,
so she believes it's better for them to die

together. A coroner said to me the other night at dinner: "Mothers kill small children; fathers kill older children." The fathers eliminate adult children because they may have a crisis when dealing with a build-up of arguments, money, misunderstandings, a deterioration of the relationship which – along with an ageing brain – lead to the loss of certain safety mechanisms. Careful though: it's unlikely that a deeply depressed person will kill and then commit suicide; it's always a subject prey to a mixed state, where depression is associated to alternating anger, agitation, aggression and sometimes psychosis.

"Generally,
one does not
kill oneself
in a rush of
reasonableness."
Voltaire

E.G.

There is a lovely quote that says: "No one commits suicide because they want to die", implying that a suicide seeks to eliminate pain. In my experience, I remember that I was really exhausted, feeling a weight that oppressed me, a sense of being cut off, loneliness, a desire for oblivion, not wanting to see anyone, talk, eat, sleep... It's true that you want to stop this darkness, you don't want to kill yourself, just stop the suffering... But let's change the scenario. I watched dramatic videos of Italian DJ Fabo, who became blind and quadriplegic following a car accident (he had bent down to pick up his cell phone), then in 2017 opted for assisted suicide in Switzerland. He wasn't depressed. He did not commit suicide because of depression. We're talking about something else entirely, of being incurable, extreme physical impairment that leads you to decide to end it. Fabo certainly wasn't the first case. Do you think we have the right to be helped and assisted in certain circumstances? If so, when?

M.S.

Yes, I think we do have the right, even if the law in many countries says not.
We all wonder who we'd ask to help us when we might be distraught with such unbearable pain, for example if we were paralyzed and surrounded by people who insisted on keeping us alive. Today, in many countries, there is a legislative barrier that prevents those who love us from helping us to overcome suffering even when, after years and years, therapy has not produced results or when therapy doesn't even exist.

"If the body is no longer able to perform its functions, is it not better to free the soul from suffering? And perhaps we should act promptly because when the time comes, we may be incapable of taking action: living very badly is worse than dying younger."

Seneca

PART 5

A psychiatrist's life

E.G.

Professor Savino, you were born in Italy, in a town called San Severo, down in Puglia, in the same building and at the same time as a famous Italian cartoonist, Andrea Pazienza. How was your childhood?

M.S.

Happy!

E.G.

So why this calling to deal with "crazy people"?

M.S.

I wanted to be an illustrator, like Andrea. Except that when we were little, he was already drawing incredible things, while all I managed were simple sketches.
In short, I was convinced that I wasn't

good enough. My father has worked all his life as a biologist, in his clinical analysis laboratory, my great-grandfather was a physician, and a cardiologist friend was also invited to family reunions... So, life conspired to have me study medicine, and at the time I saw this as a second choice, of course. I enrolled at Pisa University, where I attended a lecture given by Professor Cassano, who spoke in person of anxiety reactions. His approach was very hands-on and far removed from the prevailing concepts of psychiatry of that time. From his words, I got the impression that there was an opportunity for a new approach to psychiatry, with a more "medical" slant. These were the early years, when neurobiological foundations were discussed, as were effective drug therapies, how the different pharmacological and physical therapies acted, and so on. That lecture was enlightening, so I began to study with the intention of joining his ward. And here we are.

E.G.
In more than thirty years in the profession, have there been patients who didn't make it?

M.S.
Only a few, thank goodness.

E.G.
Do you consider them a personal failure?

M.S.
Let's say it has been painful but at the end of the day, a good physician can't save the world and no more than a poor doctor will kill everyone. A cardiologist can't stop

every heart attack in their patients.
The "failures" are less than ten. It has
been rare.

E.G.
Can you tell us about them?

M.S.
Well, a businessman who had everything
to be happy, with no money problems,
suffered from recurrent bipolar depression. There were moments of euphoria
alternating with seasonal episodes of depression that I wasn't always able to prevent with stabilizers.

At one point he was working on a big deal
that later fell through. The depressive
phase deteriorated so I recommended a
stay in a specialized clinic.He accepted

and seemed to be getting better, so he went home. But he then committed suicide with a gun he kept in his office.

E.G.
Did he tell you he had a firearm?

M.S.
No, but he died after surgery, which may have been mismanaged; it was a small bullet that hadn't fragmented and hadn't affected vital organs.

E.G.
Do you remember the case of the Italian singer Gino Paoli, who still has a bullet in his pericardium? He tried to commit suicide with a lightweight gun and the bullet lodged in an area where it did no damage, so the surgeons decided it was better to leave well alone.

M.S.

No, I didn't know. So that's why his x-rays show this bullet...He's a lucky man.

E.G.

Other "failures" you remember?

M.S.

Almost every time these people were sent to clinics because their conditions were getting worse and self-harm had increased. They were admitted to hospital, except for one person, who was referred by facility doctors to the day hospital. Identifying the risk and sending patients to the hospital facilities makes me less sad, but I am saddened nonetheless... Some psychiatrists suffer a real shock following the suicide of one of their patients. But it's part of our job.

We can't achieve complete success in preventing these tragedies.

E.G.
What kind of symptoms did you note in those other patients who didn't make it?

M.S.
I remember a merchant in a mixed state, using cocaine. I treated him at a time when he seemed slightly depressed, then he disappeared for a while, returning in a state of great agitation, even interrupting the appointment and running away.
He then committed suicide, but there was no way of intervening: it was something really sudden.

E.G.

But did you understand he was at risk?

M.S.

Well, he was clearly at risk, overexcited, but there was no way to save him, stop him, have him sectioned because he disappeared so fast and never went home.

E.G.

Other unlucky patients?

M.S.

A manager at the Teatro alla Scala in Milan, a woman resistant to treatment who was sent to a Milanese hospital where they decided not to admit her. She was offered day hospital therapy so they could keep an eye on her. I don't know how she managed to take her own life because I

didn't hear from her again. Then, I remember a schizophrenic who was treated for twenty years. Later he set himself
alight and didn't survive. I'd had him in
treatment until just a few years before, he
was doing quite well, then I don't know
what happened, but this dramatic gesture
came about. Perhaps he stopped treatment, or it had been changed.

E.G.
How many patients have there been in your
career? A thousand?

M.S.
I can't say for sure, more than a thousand,
because I started working very young,
both in hospital and in clinics.

E.G.

So, it's mainly men who slip through the net...

M.S.

Yes... But I also remember a lady with anxiety and depression who jumped from a window. There was poor response to drugs and the situation was serious, so I had her hospitalized in a psychiatric unit. I went to see her. It all happened so fast: the assessment of the risk, hospital-ization, and suicide as soon as she went home... There were external causes relat-ed to her husband's job as he was often on the road. The family didn't provide a true stable refuge, she was dissatisfied and disappointed... The aggravating fac-tor was the simultaneous presence of a bi-polar disorder and overwhelming anxiety attacks that heightened her mixed state, which initially had only been present at subclinical level. Then I met her husband, he needed an explanation and together we defined some likely causes, and we found there had also been other suicides in her family... Perhaps we could change the subject?

E.G.
So, tell me about a "miracle".

M.S.
The one I remember with greatest pleasure is the wife of a music producer who had suffered from bipolar disorder with terrible depression for years. She came to me after trying everything: Switzerland, Germany (where she'd been cheated with false tests). She could barely speak, in a state of deep depression alternating with

what were – by then – remote episodes of euphoria. She seemed hopeless. I added a little lithium to her existing therapy and after a few weeks she was much better.

Another patient came to me after throwing themselves from a window in a suicide attempt – it was a miracle she survived. She suffered from an obsessive disorder associated with bipolar symptoms. Again, with lithium and a drug to control obsession, the therapy began to bring benefits. That was about twenty years ago, she hasn't relapsed, and I still see her.

E.G.
As far as lithium is concerned, when was its importance understood? Is it a fundamental discovery for psychiatry, like ECT?

M.S.

Psychiatric medication is only a recent development, dating back to the
1940s–1950s. The positive effect of lithium on patients with schizophrenic and
manic symptoms was noticed immediately. Lithium has no real rival and is
unique in that it works quite well both in
euphoria and depression episodes. It can
be stopped and restarted but will require
a little more time to start working again,
which led me to think it might not be effective if repeated after an interruption,
but it's a matter of patience. When prescribed on the basis of an exact diagnosis,
it's dangerous to stop taking it because in
such cases there may be a manic or depressive relapse, and quite a high suicide
risk. In the right doses and used appro-
priately, it's practical, with low intoxica-

tion risks, while bloodwork and medical monitoring help avoid possible side effects. Anyone who hears about lithium is often concerned about lots of blood tests, but that's not the case. A test every so often is enough, with checks of kidneys and thyroid that may be affected. Lithium acts quite quickly, sometimes changing the clinical situation in a matter of days or weeks. The extraordinary thing is that those who respond to lithium often don't have relapses and are well for the rest of their lives.

E.G.

It's the star of medication...

M.S.

No, we still haven't found the star
of medication. For example, if you're an
artist, lithium can cause problems because it can reduce creativity.
Also, a real "star of medication" shouldn't
make you fat, cause sexual problems,
have upsetting side effects; it should always work and heal the disease in a short
time, so the patient can stop using the
drug. Because we all want to "make it on
our own". We find it hard to accept that
we're not autonomous.

PART 6

Suicide
in animals

Termites carry outright suicide devices and will detonate themselves to defend the termite mound, giving off a poisonous substance that keeps predators away. Nonetheless, while we know Hector knows he's defending Troy, we don't know if termites have the same aware-ness. If they do, then we should consider it not so much suicide but a heroic act, a form of martyrdom.

Does a dog let itself die because its owner has gone or because it's been taken to the pound? Does it know the consequences of malnutri-tion? Does it know that if it doesn't eat, it will die? In some ways, at an instinctive level, it probably does. But instinct is quite different to awareness.

Flavio Giardinelli, director of a veterinary clinic on the outskirts of Milan, explains: "I'd graduated four or five years before and the most recent animal psychology discoveries weren't available then. The owner of an automotive dump site kept his dogs chained up through the day and released them at night.

One of the dogs was run over by a car and its hind legs paralyzed, so it dragged itself along. It was prehistory, we had no CAT scans or MRI or spinal surgery. Then the dog disappeared, and they found him in quite a deep canal where there was always some water, but at that time was strangely empty. My first thought was that the dog had jumped down, that it hadn't just fallen in. I wondered if the dog was aware it was going to die. He was paralyzed, so he wanted to end it all? Even now we don't have the knowledge, we don't have the tools to decide. We can't say whether there's awareness or not. We talk about instinct, both for termites and for dogs. While we can exclude the infamous "scorpion suicide", when the insect is surrounded by flames and it contracts due to the heat, sometimes stinging itself, but in a totally involuntary manner.

In Scotland there's the famous Overtoun Bridge, the scene of several "dog suicides", but the dogs actually die when they leap towards the mink dens below. Old lions leave the pack and go to die alone in the bush. "But," explains Giardinelli, "they go to avoid being killed, because it's the weakest, so it seeks a refuge. The pack doesn't follow him because it chooses to take care of the young. Old or sick animals leave the flock because they don't want to be a burden."

There are no stories or discoveries of conscious suicides among animals, even chimpanzees. Depression is frequent, however, and apathy and loss of appetite often lead to death. But we can't speak of suicide until we can show that animals are able to understand and foresee the consequences of their behaviour. Not to put too fine a point on it.

Jonathan Safran Foer says we don't need to treat animals like humans but with respect.

Depressive syndromes in animals respond well to drugs and psychotherapies. Mario Savino says: "One of my dogs, an American Staffordshire, was very shy in the presence of humans, so I took him to a famous breeder, who is an expert in dog psychology. After a few fruitless meetings, I began giving the dog Prozac, without saying anything to the breeder. In a few weeks, the expert told me the dog was making fantastic progress, thanks to his psychotherapy of course..."

Giardinelli concludes: "I think that suicide is exclusively a human act."

Deep brain stimulation and other machinery

We met with Giuseppe Fazzari, an Italian psychiatrist specializing in brain stimulation techniques using dedicated devices.

"Stopping suicides is my mission," he says with a smile. "In my mother's family there was a suicide in every generation. I remember when I was five years old, I saw my mother confined to bed, never getting up, so now when I see a catatonic depressed subject get up and walk, I am a happy man. I get goose bumps every time. I think my childhood weighed heavily on me. I remember the anguish I felt as a boy when I heard some of the stories, which then had a huge impact and encouraged me to take up arms against suicide when I grew up. It's a tough battle and it strikes me that in many countries the most important tools for preventing suicide are not used. For example, ECT, which was invented in Italy in 1938, yet is snubbed here, while worldwide it is used for between one and a half to two million treatments a year. It's considered a jewel of medical treatment and it may not have won the Nobel Prize, but it was something that changed psychiatry and the lives of very many people."

E.G.
Just as some countries snub lithium...

G.F.
There are documents from the 1300s that talk about the ramification of lithium... Another fundamental instrument is very low-dose clozapine, which many physicians no longer use because they say it's a nuisance to take a weekly blood sample for eighteen weeks.
So, we're worrying about doing bloodwork

on a person who wants to die, whose life is in danger?

E.G.
Are you disappointed with your less enterprising colleagues?

G.F.
There's a sort of defensive attitude I see in some psychiatrists who simply say that a patient at a very high risk of suicide has a "personality disorder". I heard a patient of mine who threw herself from the fourth floor and smashed her pelvis described as having a "histrionic disorder". She was a nurse whose husband had died, and she was left with small children, in a very difficult situation. She attempted suicide four more times before I met her by chance because she was a lay nun in Ecuador

where a friend of my partner lives. She was taking lithium but hadn't had blood-work done for two years. I told her doctor and she was a psychiatrist she decided to take the patient off it... Incredible. I finally managed to get them to do some ECT. She has now remarried and lives happily, her children back with their mother and after being fostered to her brother for some time. I met her while she was out walking, and we hugged.

E.G.

Do you have more stories?

G.F.

A married factory worker was unable to have children, despite numerous attempts at insemination. She also volunteered in the police force. One day she sent a message to her colleagues saying "Just wanted to say goodbye. You're all lovely, but I've had enough." I don't know how they managed to find her in the woods where she was going ahead with taking her life. I admitted her to hospital, and she tried to kill herself with a bedsheet during her time there, so she was closely monitored. During a meeting with the great psychiatrist Athanasios Koukopoulos, I heard for the first time about low-dose clozapine for patients at high suicide risk. I phoned the ward and instructed them to give the patient clozapine immediately.

For years she sent me a message every day telling me how grateful she was and that she loved me.

E.G.

Not all stories have happy endings, I suppose...

G.F.

I recall another lady, who had been hospitalized for three years for depression, but
really, she was in mixed state, like almost
everyone at high suicide risk. Now, this
young woman, the wife of a veterinarian
and with four children, was entrusted to
me by a colleague who was at a loss on
how to help her. I started an ECT cycle
and she responded well, so she left the
hospital and went home to her children.

She had a relapse after three years of
normal life, so we repeated the cycle and
she was well for another two years. Then
another relapse, and the chief clinician
refused to authorize any more treatment.
Since then, the patient has been living in
a community and away from her children.
It's been six years.

E.G.

In Italy some psychiatrists say:
"We invented ECT. We're not stupid.
If it worked, we'd use it."

G.F.

So why does tiny Denmark, with four million inhabitants, have twelve facilities,
with one even in Greenland? There are
sixty in Sweden. In Goteborg, when a suicide risk subject arrives, first they perform ECT, then they collect case notes.
In Germany, where ECT was abandoned
for years after the Nazi regime, rejecting
anything considered invasive, today every
university city has an ECT service; in the
UK, a general hospital won't be accred-
ited unless it employs psychiatrists with
ECT skills. Because when a suicide risk is
high, electroshock is the best treatment.

It's a shame that some countries struggle to admit it.

E.G.

Mario Savino and I discussed the issue of the social perils of potential suicides. Another reason why prevention assumes great importance. Have you ever had a case involving other people?

G.F.

Yes, of course. A lady I met about twenty years ago: a beautiful woman who then developed a very serious mood disorder. She lived in a small building and, unfortunately, one day she decided to kill herself with gas. She woke up on the floor outside her home, still on the mattress where she'd been lying. She was alive but four people from the block of flats died.

From there she was sectioned in a community where there were nine male inmates: at night she barricaded the door with a wardrobe and cabinets to avoid being raped. I was assigned to treat her, and I gave her lithium with a small dose of 25-50 milligrams of clozapine. She put on some weight but lived for twenty years, dying of natural illness.

E.G.
Besides ECT there are other external therapies, complementary to drugs.
Can you list them?

G.F.
TMS - Transcranial Magnetic Stimulation - dates back to 1995. Deep TMS was approved by the Food and Drug Administration in 2013. By "deep", we mean that

instead of stimulating up to a maximum of 1.5cm below the scalp, this stimulation reaches a depth of 6cm.

Another world. Besides which, it doesn't cause memory problems. Indeed, in Israel the deep TMS appliance is recommended for treatment of early-stage Alzheimer, so it even assists with cognitive skills. And requires no anaesthesia. An extremely versatile machine.

Relatively new and important is DBS, Deep Brain Stimulation, a system of various electrodes permanently implanted to send impulses to the brain. In Italy, unfortunately, it is still used very little and sometimes badly.

Then there is direct current therapy. International guidelines don't accept it as effective with any certainty. What is it? Basically, it is a battery for electrical stimulation. How does it differ from ECT? Significantly. ECT is a deep stimulation of the whole brain caused by a seizure. Direct current is simply a shock. Which can still be scary. The patient wears a hood, two sponges are applied to the forehead – we cut dish sponges into four – so we can attach two electrodes. Then we turn it on and that's it. It can be a ritual: a regular daily treatment. Like when we go to a nutritionist at fixed appointments, we have a better chance of losing weight.

Now, I don't know if the results obtained with direct current are due to stimulation or are a psychological effect. Because there are sometimes results, even if they aren't comparable with those obtained with TMS or ECT.

Finally, there are vagus nerve therapies, consisting of retrograde stimulation of the left vagus nerve, because the right side contains cardiac fibres. This therapy is said to activate certain areas that are important for treating depressive disorder. The percentages achieved have never been significant and for a while it raised interest because it was unobtrusive.
I had a patient who has been well for four or five years. But after the device was decommissioned and replaced, he no longer responded to treatment.

Perhaps it was a problem with the machine. I changed his treatment.

There are now so many tools for prevention but it's imperative that we develop
neurostimulation culture, so it has its own
set of results, its own logic, and we can
avoid inappropriate applications.

From guilt complex to eating disorders

We meet Mario Miniati, an Italian psychother-
apist, to talk about the techniques that can
prevent risk of suicide or possible relapses.

M.M.

If we start from depressive disorder, with
its many symptoms, including sadness,
inability to experience pleasure, we often
encounter guilt.

E.G.

Have you had cases of suicide among
your patients?

M.M.

Thank heavens no even if I been in the
profession for twenty-five years.

E.G.

Have you had patients at risk of suicide?

M.M.

Well, all of us who work in the field have those and continue to have them. There are guidelines to be strictly observed. While it's true that we must adapt actions to the traits of each patient and there is personalized therapy, there are guide-lines to apply because this sort of risk is no different from severe coronary artery disease, where the patient may die. In our profession, the equivalent risk is suicide and when it's assessed as being a tangible risk, we have to comply with international guidelines – we can't afford to be hit and miss. There are psychotherapy patients who are not at risk of suicide but when the risk is behind us, and we're in a phase of even partial recovery, then we can work towards prevention, or for interpersonal cognitive restructuring.

E.G.

How does interpersonal therapy work? Does it restore social interaction to the patient?

M.M.

It's a technique aimed at reducing symptoms. Interpersonal therapy is structured in three parts: initial, central, final, for about twelve sessions. Four problem areas are considered: two are interpersonal deficits and grief; the other two are role transition and interpersonal conflicts, which occur more frequently in younger subjects. An example of a role transition? Losing your job or retiring will sometimes lead to depression and a suicidal reaction can appear if we add organic and physical pathologies to the mix, as well as poly-therapy that may bring depressogenic effects, taking into account an interpersonal deficit component.

At that point a series of specific techniques and strategies on how to reduce depressive symptoms are assessed. There will also be some "homework", to reactivate a social component if possible.

"Those who choose suicide have encountered a shattered mirror, they can no longer recognize themselves in anything. They have been stripped of their image."

Massimo Recalcati

E.G.

Do you suggest actions to stop the patient feeling alone?

M.M.

Let's say that the therapy also seeks to strengthen individual resources and we can agree on a set of exercises for a patient to put in place to reactivate social interaction and a valid interpersonal support network. Of course, on its own this won't solve the problem but it's one of the elements that alleviates symptoms of depression in quite an intense way.

To be in a state of depression all day long is quite different when interacting and feeling useful again, performing a series of functions. It's one thing to watch TV all day at home – not even watching just staring at the screen – and quite another to go out and do something. Depressive symptoms affect the quality of interpersonal relationships. As the quality of the interpersonal relationship deteriorates, the symptoms become stronger and we have to break this vicious circle. We can intervene on physical symptoms with drugs while with psychotherapy we can influence cognitive and interpersonal restructuring.

Clearly, for the elderly this can be more difficult than in other age groups since the interpersonal deficit is much more intense. For example, in role transition, the subject will find themselves dealing with a series of tasks that a life partner would undertake, and not knowing what to do, they see themselves in a passive role compared to the previous active social role.

E.G.

Let's not forget that if lose your job or you retire, you also lose a group of colleagues. I'm not saying they were a family, but still... You could say there's a "loss of camaraderie": the boss who chatted about a daughter's broken heart, the warehouseman who won a lottery... Stories that filled a day

M.M.

Yes, that also has an impact.

E.G.

Another massive front is encountered in eating disorders, especially in young women.

M.M.

True. There are stories of young women whose BMI falls very low: 1.70 metres in height and weighing only twenty-five kilos. But they can recover perfectly and go on to lead a normal life and start a family, after having been on the edge of the precipice in their daily wager with death.

Conclusions

We've seen that in the majority of cases, there are a number of conditions in developing suicidal risk:

1) A subject will have a predisposition, in other words suicide is not something that can happen to anyone.
2) One or more prolonged stress factors must be added to the predisposition, perhaps combined with a guilt complex, triggering a depressive disorder.
3) The depressive disorder must be mixed state, namely depression accompanied by anxiety and agitation. The depressed person who stays in bed or sits in front of the TV, is not at great risk. The act of suicide requires physical and mental energy.
4) A wrong diagnosis, underestimating risk, especially when the disorder begins.

This is what we've discovered so far and it's a new starting point for trying to reverse a trend.

The latest World Health Organization estimates predict an increase in suicide victims in coming years, with on average one every twenty-one seconds, and a suicide attempt every ninety seconds. A phenomenon we can't ignore any longer and it must be addressed across the board, without fear or prejudice.

We produced this publication precisely to break down stigma and clichés like "there's no hope, people at risk of suicide always will be". This is untrue: they are at risk only for a limited period of time. Or, "talking about suicide can trigger suicidal behaviour", which is also untrue as an open discussion on the topic helps the person in need to ask for help and often provides relief and understanding.

"There are invisible suicides. We remain alive through pure diplomacy, we drink, we eat, we walk. The others always believe in us, but we know they are wrong. We know we are dead."

Gesualdo Bufalino

Essential bibliography

Temperament profiles in physicians, lawyers, managers, industrialists, architects, journalists, and artists: a study in psychiatric outpatients. J Affect Disord. 2005; 85 (1-2):2016 (ISSN: 0165-0327); Akiskal K.K.; Savino M.; Akiskal H.S.

Genomewide significant regions in 43 Utah high-risk families implicate multiple genes involved in risk for completed suicide, Molecular Psychiatry, Gray D.and others.

The close link between suicide attempts and mixed (bipolar) depression: implications for suicide prevention. J Affect Disord, April 2006; 91(2-3):133-8. Akiskal H.S. and others. Suicide in a large population of former psychiatric patients. Article in Psychiatry and Clinical Neurosciences 65(3):286-95, April 2011. Koukopoulos A., Tondo L. and others.

Help With Postpartum Depression, American Psychiatric Association, www.psychiatry.org.

NIMH Answers Questions About Suicide, www.nimh.nih.gov.

Preventing Suicide A Resource for Media Professionals, Department of Mental Health and Substance Abuse World Health Organization, www.who.int.

Final Exit, Humphry D., Dell, 1991.